How Thi...

by He...

Table of Contents

Consultant:
Adria F. Klein, Ph.D.
California State University, San Bernardino

capstone
classroom
Heinemann Raintree • Red Brick Learning
division of Capstone

Simple Machines

Machines make work easier.
Some machines have many parts.

Simple machines have few parts.
Simple machines make work easier
by helping us push, pull, or lift things.

Pulleys

Pulleys are simple machines.
Pulleys use wheels and a rope
to raise, lower or move heavy things.

We use pulleys to raise a flag
on a pole. We also use pulleys
when we open window blinds.

Levers

Levers are simple machines.
A lever is a stiff bar that rests
on a support to lift or move things.

A stapler and a scissors are levers.
A seesaw is a lever too.

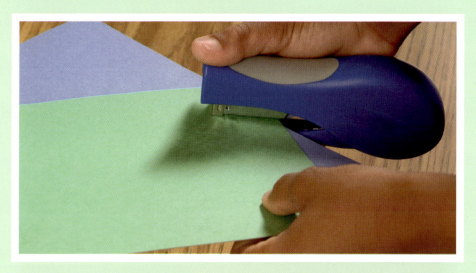

Inclined Planes

Inclined planes are simple machines.
An inclined plane is a slanting surface
that connects a lower level to
a higher level.

Ladders are inclined planes.
Slides are inclined planes too.

Wedges

Wedges are simple machines.
A wedge is an object with one thick
edge and one thin edge that cuts
or stops something.

Axes and knives are wedges.
Your teeth are wedges too.

Wheels and Axles

Wheels and axles are simple machines. A rod called an axle goes through the middle of a wheel. Together they help lift or move loads.

Rolling pins are wheels and axles.
Wheelchairs are wheels and axles too.

Screws

Screws are simple machines.
A screw is a fastener with grooves
and spiral threads. Screws hold
materials together.

A lightbulb has a screw. The lid on a jar is a screw too.

How have you used a simple machine today?